To world literature

and toward a time when

ultimate truth and enlightenment

may become more deeply understood and sought.

Tabula Rasa

On the manifest need for fundamental philosophical redirection

Published by The Logos Foundation
Post Office Box 758
Carson City, Nevada 89702

Printed in the
United States of America

First Edition

Designed and set in ITC Korinna by
Feiner Professional Support
Post Office Box 7351
Laguna Niguel, California 92677

Library of Congress Cataloging-in-Publication Data
SOMOS, date.
Tabula rasa : on the manifest need for fundamental philosophical redirection / SOMOS. -- 1st ed.
p. cm.
Includes bibliographical references.
ISBN 0-9623445-0-8 : 8.50 (est.)
1. Philosophy--Miscellanea. I. Title.
BF1999.S5728 1989
191--dc20 89-80842
CIP

A consistent, comprehensive,

rational and consequent worldview

and universal perspective.

PROLOGUE

The cover of this publication is blank, pure – Tabula Rasa.

That is the principle by which we want to set out and conduct our inquiry into the deeper questions of existence. No preconceptions, no prior notions, deceptions, superstitions, archaic traditions, ignorance, no hypocrisy, no spoils – just the facts, the real, reasonable, rational facts as we have come to know, comprehend and understand them, with our best faculties applied.

We have come a long way since realizing that the earth is not the center of the universe. But now we have got to finally go on and separate *all fiction from fact,* true reality from all that stuff of myth, old tradition and of the (unreasoned) imagination, and come to grips with the like realization of the way *all things really are, fundamentally,* and go forth in fuller enlightenment.

We have had the flat earth people, the "it-all-came-about-in-seven-days" people, the Santa Claus story, and the "believe-it-or-go-to-hell" groups. Not to speak of the more ancient God figures of Egypt, the fledgling thinkers of Greece, and the Nirvanas. But we are now hundreds, thousands of years after all that. We ought to be able to recognize, define and understand the true nature of things a lot clearer by now.

So let's see where hard, reasoned fact bucks up solidly against all that mushy, murky world of the made-to-believe.

In the end, our great surprise will be in the simplicity of it all.

* * * * *

REGISTRY OF THOUGHTS

Prologue **11**

Author's Comment **17**

To You – The Reader **21**

Introduction - *where we stand* - **25**

On the Ultimate Essential, Consequential 31

On Relative "Reality," Realization and True Awareness 34

On the Nature of Things and On Essential Nature 40

We are All Sinners 45

On the God Concept 48

On the Physicist's Equation 50

On Subconscious Awareness 54

On Original Thought, Ultimate Knowledge and Understanding 57

On the Province of Human Judgment, Horizon and Philosophy 62

Registry of Thoughts *(Continued)*

On Imperfection and the Perfect State 68

On Our Time 71

On the True Maker - Deity or Impostor? 76

On the Condition of Religions 78

On Man's Shame 80

On Eternity and the Temporal 82

On the Necessity and Esoterica of the Inquiry 84

Reflections 89

On Nature and the Object World 96

On the Categorical Judgment on the Non-cognizant Worlds 98

Concluding Remarks 101

AUTHOR'S COMMENT

This publication is not to be critically acclaimed or critically condemned. It is simply an effort to help recognize, understand, define, state – fundamentally – the actual, true nature of things and of ourselves.

You may contribute by defining, stating things better, the way *you* perceive them. But for heaven's sake, do not just sit there and criticize this effort. Instead, contribute a better effort! But remember, we are in the arena of philosophical inquiry. We are not into science, religion, or whatever one's bag may be. We are into neutral, philosophical inquiry.

What do you think, fundamentally, this Universe is all about?

That is the question. Again, please do not respond that Allah is great, that God is almighty, or that you do or do not believe in a creator, and that when it is all over, we are going to be in Heaven (or in Hell, as the case may be). Or that eventually, we shall be able, through science, to know all there is to know about the Universe. For if we knew it all at this very moment, even with all that knowledge, we still would not have answered the *fundamental question of what this Universe is all about.*

So far, I think all we can know and conclude is that this Universe is about *SOMETHING -- BUT WHAT?*

Our ultimate fulfillment is not in our better knowledge of technology, in a more developed, "modern" life, or in adherence to one of the many intangible belief systems handed down to us – *it is in our coming to truly understand the nature of that certain SOMETHING, in our basic realization and ultimate transcendence of ourselves*, of our wanting essential nature, of our very creature life, through the recognition and acceptance of actual, real truth principles.

Why do some of us ask these questions?

Why do some of us go to church?

Why are some of us hardened, uncaring criminals?

And why, oh, why, do most of us just live a humdrum, "content," yet unfulfilled, fearful, uncertain life – not asking, not knowing, not understanding – seemingly not caring about much, other than their own immediate little selves and self-interests?

There are answers to these questions, and try we shall to wire the reader for some comprehension and understanding of them.

"Why wilt Thou distance Thyself from Us All
And from our Opinion?" --

I write not to please you, but that you should learn something.

--- Goethe

TO YOU – THE READER:

Do you see the need for fundamental philosophical redirection manifest?

For if you do not, if you are happy and content with the way you are and see things and feel no need for change, no need for our deliverance, then we have very little to talk about, actually. You can go back to your children, your church, your job, your national purpose, your physical world, procreation, and continue your human endeavor on the familiar circular path.

Never even mind our long, sad history; just consider that in this our time alone, fallacious, imperfect, wanting human aims, interpretation and understanding have brought on conflagrations with our entire world at war against itself. And no sooner a conflict "settled," humanity has raced on to amassing ever more deadly weapons, supposedly to "settle" yet the next conflict. Only by now, we may just "settle all," right into complete annihilation and oblivion. Oh, what full proof of basic, very basic wrong premise in human pursuit.

In a sense, our physical oblivion is pre-ordained alright, but it better be through higher awareness, to the upside, not for the sake of lowly, greedy destruction and basically physical pursuits. For if we are not led out of the yoke of the physical world with "graduation papers," with higher spiritual awareness, our lower energy will surely have to stay with, or return to, another imperfect state to learn the proper way up and out. No way could we reach a high, non-physical, spiritual level of existence with low and wrong impulses in our hearts and subconscious souls. That has already happened, hasn't it? It is happening right now, isn't it? For we are here, in our imperfect state; low spirit created this, us, the physical, the temporal, before! How simple, how easy it would become if we only recognized our problem, if only we were able and willing to adopt, and follow through consequentially on those higher principles of true reality – that of infinity, of eternity.

Material, physical motivation is not at the core of our needed redirection. To be sure, righteousness, security and economic and social considerations and order have their place as we prepare for our way home, but our real and highest values lie in the pursuit of our true destiny: the achievement of deepest spiritual maturity and awareness in our subconscious – and the resultant tangible transformation to a higher essence of existence.

" . . . and whoever strives for perfection strives for something divine."

--- Michelangelo

"Certainly, it is

heaven on earth

to have one's mind

move in charity,

rest in providence, and

turn upon the poles of truth."

--- Francis Bacon

INTRODUCTION

- where we stand -

When it comes to our real destiny, we still perceive much like the wild tribe in the jungle that is waiting on top of the hill, by their bamboo-built airplane replica, for another airplane to arrive, except we are using for props in our "civilized, cultured" pursuits our varied religious memorabilia instead.

In the case of the tribe, once there had actually been a "real" airplane coming from the heavens, dropping food and supplies on their starved people. Their comprehension, or awareness level, though, could not fathom the realization that fellow human beings were actually behind their "miracle," so they continued "believing" in what they could muster – that some abstract "God" had come to them from the heavens with help. And now the hungry lot hope and pray on the hill every day for their "God's" return.

Well, we know better, don't we? --

Evidently, we don't.

In our own ways, and perhaps hypocritically, we hang on to our old beliefs that are simply no longer relevant and in sync with our comprehension level. In the process something, perhaps more than we realize and are willing to admit, is going awry. Public morals, ethics, are no longer guided by a fully accepted belief system, and there is nothing tangible to replace the old values. We have not seriously considered alternatives, status quo establishments are comfortable saying "yes, yes" to the old and traditional, and we follow merrily along in the manifestly out-comprehended old ways.

No matter how noble, how ancient and durable our varied beliefs may be and may have been, they smack of handed-down tradition, status quo, control, blind acceptance, superstition and error, even misguidance and deceit.

Time has come to go with the blessings of our higher awareness and comprehension and to seek a rebirth of our beliefs, in holy matrimony – in sync – with our new knowledge and understanding.

We must probe not only the visible, physical form as scientists would do, but especially that which is beyond touch and sight. For we must replace our old belief systems with new truth values, shed of any and all superstitions, tales, deceptions or "sacred," fallacious hand-downs. Remember Galileo and the Inquisition?

We may know ever so much about many things –

but if we fail to understand our own essential nature,

then we cannot truly understand much of anything.

Tabula Rasa

ON THE ULTIMATE ESSENTIAL, CONSEQUENTIAL

In search of the ultimate criterion, there are mighty few disciplines to go through, really. There is Science, Religion, Philosophy (thinking, reasoning, deducing, concluding); there is Hocus-Pocus, and . . . can you think of *ANY* other discipline that might cloister life's ultimate secrets, purpose and promise?

We could start by deliberating never-ending scientific considerations, or going into long religious dissertations, but let's first take an overview of what we are trying to accomplish. In ultimate ends, do we want to find science's deep-seated last secrets, go with near-perfect delusion and edicts of religious rites and traditions, or is it more appropriate and important to find whether all is, *very basically*, rational, right or wrong, good or evil, adequate or inadequate, in fundamental nature? I opt for the latter.

If all of science were not to measure up to "good, adequate, right," then we would surely have to ignore it as a means of exploring the ultimate mysteries of existence. By proper, correct judgment, most assuredly the ultimate good is paramount – a prerequisite – in any consideration on the adequacy or rationality of being and existence. Is the ultimate criterion, then, an ethical, moral question? A resounding *YES* – and with the proper awareness it can just as well be a rational one.

Shall we go fishing on the lake this afternoon, or shall we go hunting in the woods? Is it a moral, ethical question? No, not for most. Instead, it is likely a question of preference, mood, choice – of rationale only. But now ask the same question of a total vegetarian. Then it becomes a moral question, not one of choice or preference. Rationally, reason might tell the vegetarian it would be more fun on the lake than to tramp through the bushes hunting. But to him, it cannot be a question of choice, preference or rationality in choosing between one or the other. It is, instead, an engulfing, overwhelming matter of conscience either way: killing another creature for his own physical maintenance -- repugnant! The moral question, then, determines if he should engage in either activity at all; if he needs in fact any of science's knowledge, say, for the fishing gear, or the hunting guns. Having decided beforehand that he has no use, no necessity, for either – he can, in fact *must do* completely without the heavy hand of science in reaching his ultimate objective.

By rights, then, *in ultimate consequence, right or wrong, good or evil, adequacy versus inadequacy, becomes the final determinant.* The circle becomes complete when morality is in sync with rationality. Moral and rational, the correct combination, the way it should be, and is, when things are right.

It is my special, sure sense that it is this simple way with (our) life's last consequence – the ultimate criterion being a question of the morality, the rationality, the adequacy or inadequacy of the nature of our being in light of highest universal principles – as the brief encounter with consciousness that is human life struggles toward its full self-realization.

ON RELATIVE "REALITY," REALIZATION AND TRUE AWARENESS

In this movie set, the two-man raft, trying to land on a remote island during a severe storm, is being rocked back and forth between the boulders near the landing site . . . the two occupants have just confided to one another that they cannot swim . . . the excitement mounts. Will they make it, or will they crash against the nearest boulder during the next oncoming wave . . . and perish?

"The cameraman seems to hold nice and steady on his safe little platform," I say. To which the answer comes: "You don't have any imagination, no adventure in you – you can't go through life with your heavy foot on reality at all times!"

Reality? -- Really?

It must be news to most that our actual, live reality is the most adventurous, exciting thing there is in our lives. To find true excitement, do we really need all that make-believe? Just try and gather up the live awareness, not imagined, but real, that you are travelling through space on a small sphere (huge to us) at a speed that would "blow all your imagination" if you could only see it in its true reality. You are circling around a star, and the star circles around in yet a much bigger space. And we are not sitting on a "real" – but in actuality make-believe – platform in a make-believe scene for our observations. We are sitting on the actual, life-size reality of this globe, and we have with us, or within us, a mind that can make not only real, true observations, but also deductions, reasonings and conclusions about that which we are observing.

"*That*, my dear, is true excitement, true adventure." The real experience of life, of living, of finding that which is actual, that which is ultimately true and consequential. For with every observation, every passing day, week and year, we are making new deductions and conclusions – we are getting a step closer to understanding ourselves and the meaning of the Universe.

And I am so excited about that which I have already found, and about the hard and fast knowledge that I am finding the deeper answers of this, our journey. How do I know? I can see it clearly, in my mind, and in the real world. Both are revealed to me all at once.

Now my dear cat would not know what to do with what has been said. She is very lovable, hungry at 8 in the morning, and hungry again at 6 in the evening, sleeping away half of her life -- not too different from our life, is it? At least of many, perhaps of most of us. Just a bit of a difference in lifestyle and form – a horse race or similar diversion here and there – a more or less complex way of life . . . Well, let's not overdo this comparison; after all, the cat is an animal, and people are human . . . !

But do you know something? In many, perhaps even in most cases, humans live surprisingly much the same way, for the day, for the moment, never wondering what, in true essence, it is that makes things go round – rather just satisfying their individual, mostly selfish needs, wants and desires. Surprisingly much like the cat.

Now, is this because most people do not have in them much of that which separates animal from human, or is it because most humans just don't care?

How could they not care *if they had the right awareness*, conscious or subconscious, to ask the timeless, deeper questions? Of course, as we shall see later, the foretelling truth is that in a proper predisposed higher subconscious state, life energy would not become amalgamated with physical affliction, the flesh, the way we are, to begin with.

What kind of awareness, then, are we looking for? The awareness that would make us jump away from a falling rock? No, our cat would do that, perhaps even better. An Einstein's scientific grasp, awareness of something that not even he could *prove* to himself despite his "blind" understanding and comprehension of it? Or the awareness of a conscientious objector who realizes something most others don't, and it is a compelling force to him?

Which awareness is it that matters?

> ***It is the awareness, the realization, of the true essential nature of all things and of ourselves***
> *(by which to establish a proper universal judgment and determination).*

In theory, rationally, we could spend the rest of our time on the question of the truc essential nature of things and perhaps, through such fundamental philosophical inquiry alone, eventually make it home. But in reality, it is too big a step for most to comprehend. For we are born as who and what we are – yes – in essential nature! And for most of us, the full distance cannot be mastered, our subconscious essential nature overcome, attaining enlightenment on rationality, on philosophical deduction alone; true awareness, comprehension, understanding – the proper consciousness level, are needed prerequisites to go hand in hand in reaching the light.

Our capacity for comprehension and realization of our essential nature, like of so many other things, is closely tied to what we are, deep inside us, in our subconscious being and awareness. In a simplified fashion, being who and what we are, we are what we were born as and with, and the basic crux of it lies in our capacity of being aware just what, who that is. Call it self-realization, if you will.

Curiously, many highly learned folks do not appear to have it: the awareness of the thorough inadequacy of human existence in light of highest ultimate principles – of perfectness (look for "knowledge versus cognizance, understanding" later). While even our perfectly normal body is, in essence, "a mess," it is afflicted man, unable to physically function on his own, that proves the point more dramatically. Have you ever witnessed a thorough autopsy, preferably in a state of deeper meditative contemplation?

That we do something, act, on the condition of our "mess," that is the message we get from true awareness through our conscience. All empirical existence is physical affliction in varying degrees, including our human "being." Supposedly having reached the "Nirvana" and going right on acting out and procreating its antithesis, physical life, is full proof that such a purported "Nirvana" is, in fact, mere Fata Morgana. A true Nirvana state should lead to *Paranirvana, the knowledge that you won't be back in, nor want to pass on, the flesh. Then you have in fact graduated up, you are delivered.*

Now, then, learning these principles and *developing in us that higher subconscious awareness, this is our real task, our "religion," our door to salvation and deliverance.*

The way is through initially conscious, then acquired, gained *subconscious awareness and realization*. Just keep watching, listening, observing, reasoning. Inside and outside ourselves, through constant, meditative contemplation – until the walls, or rather the depths, of our subconscious are penetrated.

After having gained the new awareness, realization and understanding and their absorption into our subconscious, we shall eventually emerge in a higher state of being, a new essential nature, much like the adult that has grown out of wanting to play the children's games of marbles or hide-and-seek.

The transformation will have taken place, you shall reaffirm and become part of the non-physical realm and reality.

ON THE NATURE OF THINGS AND ON ESSENTIAL NATURE

Most of the existing, that which is, does simply that: it just exists, it just is. It never seems to bother to define itself; it's just there.

But to be sure, everything that exists, all there is, lives, is by virtue of its very existence, its very being, a certain something. What that certain something is, innately, fundamentally, that is the essential nature of it.

What, who, then are we, essentially – in pure, ultimate essence? Mortal, physical bones, flesh, blood – or a mental, spiritual power, an energy that uses the physical container as its vehicle, its carrier? Which, in true reality, is our actual essential nature – the container, or the life, the energy that is in the container?

We have to consider the whole – we cannot just study the brain, or the digestive system, or any or all of our physical body, ad infinitum, for final answers about ourselves. Our consideration is of a much deeper character. When little birds, with brains to match, have summer and winter homes in two different hemispheres of our globe and have the wherewithal to find the same building sites over and over again thousands of miles distant where to spend their next season, without a compass, directions or any navigational aids as we know them; when a little spider constructs a masterpiece of architecture in its web, then we have to look elsewhere than in the human, supposedly know-it-all, "top-of-the-line" makeup for final answers.

Alive, living, we are obviously something different from when we are dead corpses. Dead, we are plain simple physical phenomena, comparable to the rocks and boulders in the desert. The energy in that dead body is merely mechanical, like that of the rocks, or of plutonium, etc. Thus, the difference between our creature life – our live state – and our dead state, is the essence of that which lives in us. In other words, that essence is our basic and essential being, our essential nature.

So what should concern us is that invisible "thing," that live essence which comes to the "body" at conception, and leaves it at death. Concern over why it needed to be amalgamated with that "mechanical," physical phenomenon of a body in the first place. Yes, concern over, cognizance of, that invisible thing, that true essence of our being, our innermost essential nature: the subconscious energy that transcends the physical. Simply put, our essential nature is that which is deep within us, it is our subconscious, our "invisible" self, our true life energy, our live reality. The invisible: a world in and of itself, transcending the physical "us," but in it having a temporal, if erroneous, expression and apparent manifestation.

Something missing here on the way to explaining the essential nature of all existence? Yes, like what the essential nature is of the poor limited soul of the bird that finds its winter home at the other end of the earth. Or of our resident master architect, the spider. Of our busy ants. How about the essential nature of an oak tree, perfectly grown from its acorn seed? And of a rock quarry, and of the Grand Canyon?

By now we begin to sense that philosophically, essential nature is in some contrast to purely object, passive physical phenomena, the natural elements, natural laws, to nature itself with its programmed processes. As the very name implies, essential nature denotes "essence" in being. The *"Nature of Things"* is thus set apart from *"Essential Nature"* – the latter signaling an independent strain, manifestation of existence, of life, of living energy, up to the level of self-cognizant, self-realizing, accountable being.

The basic nature of a rock – as of any object – is that it is just that: a rock, a mechanical, physical (energy) phenomenon. We might say it has in and of itself no independent essential nature or, rather, no subjectivity. It is simply a "manifestation of nature," part of a "bigger picture" if you will, without its own "will" and living, determining essence.

The fact that object water freezes at 0°C is hardly a matter of its *essential nature*. It is, rather, a process characteristic of the natural elements. Essentially, scientifically speaking, the water's components are hydrogen and oxygen (H_2O), but like most things in science, this does not explain H and O fundamentally, philosophically. As it were, the stuff that H and O are, innately, would have to explain what water is, but as a practical matter, we commonly know even less about those components individually than we know about the water they form together. So, as usual, solely scientific analysis does not take us very far by way of fundamental answers and explanation.

Purely mechanical (apparently unmotivated) energy, the passive, object world of physics, the world of the chemical elements, of non-cognizant nature, must be a separate chapter in our inquiry. The rock in the desert – or the Grand Canyon – (or the bird, or the oak tree, or fire) have no concern about religion or conscience – it therefore follows that they can only be dealt with by us philosophically, categorically, as opposed to essential being in which self-realization and self-cognizance have occurred. In the latter, the responsibility, accountability, shifts away from the "bigger picture" non-cognizant existence to the self-cognizant, self-realizing being whose philosophic quality and inclination, and resultant conscience, become the responsible agency of decision, judgment and determination.

Now all that has been said here should not be taken to mean that essential nature is lacking completely in the object, passive world, for it is still a legitimate question to ask: *What is the essential nature of the Universe – and anything in it?*

Yes, surely there is essence even in a rock, in the Grand Canyon, in the tree and, of course, in the bird. It is just that inanimate and non-cognizant existence has to be viewed categorically, judged as a complete unit – encompassing space, time, matter, all physical phenomena, the elements, non-cognizant "nature" – as if it were a helpless child unable to answer or speak for itself, and a philosophical inquiry and determination is due for it as a whole, as its "responsibility" is set apart from that of our own cognizant, aware essential nature.

Thus, the crux about it is that on the one hand, the accountability belongs to a bigger "agency" of responsibility, while on the other, we limited but cognizant, comprehending, mature beings are responsible within "*our* jurisdiction." In other words, if we know our places, we are going to get along just fine philosophically, and as a whole.

"WE ARE ALL SINNERS"

In universally acknowledged acceptance, we recognize our shortcomings, our inadequateness, but we insist upon persisting as "sinners" – yet all the while claiming that we are "the greatest," created in the image of "the perfect God."

There is a fellowship in man, of intended good will at that, in this acknowledged inadequacy – you can see it in the churches: the hand-holding, the singing, the chanting, "we are all sinners" – they know it. The contradiction is there, incontrovertibly, and that is one of the key puzzles we have to solve.

We are either of a perfect God, entitled to aspire to perfectness, to "come home" to a "perfect world," abandoning an imperfect one, or we are not of a perfect God, somehow misplaced in chaos, in a scandalous universe, condemned to an imperfect life of wanting and suffering ad infinitum, as in fact we seem to perceive and perpetuate it for ourselves.

Which shall it be?

Behold, we are on the threshold of coming into discovery, awareness greater than any other: the conscious recognition, the *knowing* cognizance of the reality of our inadequacy under that perfect God's true, highest principles. Put another way, what we are discovering are the true, highest, ultimate principles, which we identify as "God's." Contrasted against them, we can, will, and must, with proper awareness, see immediately that we fail to measure up, with or without the chanting.

Again, ultimate human fulfillment will not come through more and more complex, "comfortable" lives and unimagined gadgetries, but solely in *total spiritual maturity*. Awareness on a level that will make existing, being on this plane here a venture altogether passé, manifestly and resolutely incompatible with our deeper, enlightened insight and self-cognizance.

Man, you who pretends to be searching for the truth -- you are not, really.

Instead, all you ever do is try and justify everything for and within yourself.

ON THE GOD CONCEPT

I wish someone would explain to me just precisely *what* they mean, exactly *what* they believe in, when they say "I believe in God."

I cannot escape the feeling that they are unwittingly, unconsciously saying: I (simply must, I want to) believe in an insurance agency to "provide for me, to protect me from the (feared) unknown, to take care of my soul" – an agency to apply to for salvation and deliverance when they can't see their way clear on their own.

A great philosopher once put it a similar way: Prayer, "the begging," is asking God to do for them that which they are not prepared to do, earn, for themselves. On Catholicism, he said: "The Catholic religion is a directive to beg for entry into Heaven, which to earn for oneself, being deemed too demanding, would be too uncomfortable – with the priests acting as (instigating) intermediaries in this game."

The concept of a God is alright. As a philosopher, I do not or cannot take issue with the concept of GOD as a symbol of the perfect. I only take issue *with what we perceive God to be*, that is, with the very erroneous notions we have on the subject. God is what's right, that which is true, adequate, perfect, eternal, infinite – therefore, spiritual (non-physical). Consequently, if we want to be going in the direction of "God" and really consummate God-like being, then we have to become spiritual, as opposed to continuing on in the physical realm.

Now go ahead and argue with that!

ON THE PHYSICIST'S EQUATION

$E = MC^2$ will never go anywhere in relationship to a unified field theory because in its intent, the equation does not go back to, it does not include, ultimate origin. To say that two things are equal does not begin to explain what either "thing" is and whence it came, and all the other questions of its reason, rationale, purpose and ultimate destiny. There will never be a unified field theory "explaining everything" that does not figure in the original state and causation; in fact, I hold that unification is possible *altogether only in the original, polarity-less, perfect state.*

In actual true fact, but not as envisioned in the equation, E accounts for immensely more than just that equivalency, for Energy – through its true origin, polarity-less free energy – does reach back to the first, eternal causation. But as there is no place for such free, polarity-less energy in the physical realm, no unification is ever possible on this level.

What has to be said about the physical world, and therefore also about $E = MC^2$, physicist Einstein said best himself, stating: "What interests me is whether God could have had anything to do with it . . ." So he himself sensed, quite typical of the scientist, that the original causation, which we shall call God, was left out of the deliberations, particularly in this case, in his application as regards the equation. Still, he equated "Energy," which in true reality is of the stuff linked to original causation, God, with what only phenomenally, physically, temporally exists – Mass, times the constant in the physical world, the speed of light. Not good enough. Energy and its origin is, in a quite literal sense, *infinitely* more than just equal to that which is purely physical and temporal.

It thus follows that just simply $E = G$, Gravity, would more adequately represent Energy to our metaphysical level of comprehension. Leaving out the temporal components, the physical *(M), Mass*, and (C), the speed of light in a physical world, would certainly be a lesser shortcoming than to leave out, as in Gravity, the potential, or, rather, *certain link to the true essence of everything: the non-physical, infinite, eternal, perfect state original energy causation – call it (the link to) God, if you will.* Gravity being effective to infinity should tell us without a doubt its link to that original energy, confirming our error in the assumption that "missing matter" is our (perceived) "phantom gravity source."

To help your imagination, picture the *age* of the Grand Canyon *as we know it today* – then back up further in time *by the age of the deep rock formations themselves* before the rise of the territory creating the high plateaus around today's Canyon. Then, "juxtapose" those billions of years with the imagined distance of the "source" at the speed of light – some 12 to 15 billion *LIGHT YEARS AWAY(?)* – and picture thus the energy required at that (infinite) distance to keep the known Universe in tow! *THERE* is a (metaphysical) idea of the true magnitude, of the force, of the unimaginable *VALUE OF THE ORIGINAL ENERGY*, behind it all.

Invisible, untouchable as its mother – original, free energy – gravity obviously then looms much larger in ordering the Universe, atomically (infinitesimally) to cosmically (infinitely), than could ever be demonstrated in the equation's MC^2 in that representation of, reference to, mere physical phenomenon. By itself – without the reign of that greatest universal force and leash of universal law, Gravity – the physical universe would be hopelessly chaotic in all circumstances, certainly altogether incapable of existing as we know it.

At some point, infinity (and the infinitesimal) will transcend the bounds of physical phenomena, of our phenomenal worlds; this well beyond our province of empirical comprehension in either direction – hence our need for the *metaphysical* and unification needing to reach far beyond any one of our known disciplines, in fact craving for all of them to come together as a prerequisite to any hopes in apprehending the goal.

Therefore, rationally, not until we unify our religion, science and philosophy in one fundamental, comprehending awareness and understanding about everything, shall we come close to any real concept of field unification in our minds, and in fact. See above for the bottom line on polarity-less unification.

In man's quest to find real purpose,
there can be no doubt that he sees the
overriding mission of his existence in
the attainment of higher status than his own.

How long man is going to find that true
about himself while failing to initiate some real,
tangible steps toward his transition to
such higher level in essential nature,
is a matter of great curiosity and puzzlement.

The problem is that man insists on remaining in
his (limited) physical state of being, in "seeing,
hearing, touching, experiencing, controlling" things.
Doings that he cannot conceive the higher levels
affording him, so he keeps groveling in his imperfect,
suffering, "controlling" yet helpless, hopeless state.

Once again, the key is in the total awareness of,
and knowing faith in, the true reality of the
higher level – which man evidently still lacks in
order to undertake those tangible steps off "his pad"
and venturing on up and out of the physical realm.

ON SUBCONSCIOUS AWARENESS

We hear of meditation, of introspection, of going inside oneself. We know about consciousness and conscience, about right and wrong. Why do you think that is? And why are some of us drawn to it more than others?

We arrive on this scene on different levels, different frequencies of consciousness and awareness, you might say in different stages of perfectness. The lower our stage, the farther we are from the perfect state. Our bodies are carriers of energy – imagine it to be electric current. The higher the "frequency," the intensity, the sensitivity of that energy, the more "perfect" the being, one's essential quality.

Now, then, how do we influence that quality, the frequency of the energy in us in order to grow in our consciousness, in our (subconscious) awareness?

We have done it for centuries – through religion, morals, ethics and introspection – by listening to our reasoned judgment, to our conscience.

Religion, however, has turned onto itself, it has become dogmatized and out of touch with the real truth and therefore with our higher level of awareness. As a consequence, we are left with just loose pieces of morals and of our own conscience, without a charted course based on an overall picture. Actually, simple materialistic chaos is descending upon us as a result.

It is certainly so that every thing, every being, creature, exists on its own, different level. Among humans, it is with consciousness, awareness, comprehension commensurate with one's subconscious status.

Our mission in life is to improve that subconscious quality and awareness in us. For it is our subconscious that determines our true being: *who, what* we truly, essentially are – and are going to be – no Hocus-Pocus about that.

The learning process in this is like any other learning process. First, you must want to learn, then you dedicate yourself to learning, and when your vibes become in tune with increasing awareness, you will never want to get off.

Our inclination, our quest to learn (or not to learn) may be pre-conditioned within the space of a particular life span. If the "current" within us is on a declining cycle (as in a hardened criminal), it may be difficult to reverse that negative flow within the space of a mortal life span. But under normal circumstances, given the right conditions, environment, influence and teaching, our minds should be in a receiving mode for moving towards spiritual development and attainment of deeper awareness – first conscious, then subconscious.

ON ORIGINAL THOUGHT, ULTIMATE KNOWLEDGE AND UNDERSTANDING

The old classics did not consider Philosophy the mere "learning or recital" of others' thoughts and quotations. They originated thought, albeit fledglingly, when compared with knowledge and understanding gained since. So it is that if we want to find new, basic comprehension, we too, from our new vantage point, must go back to origins, to fundamentals, again thinking things out to their original causes. That's what they in their own early ways were doing, and that is the true meaning of philosophy.

Ever heard anyone lately, in recent so-called philosophy, dealing with those truly causative subjects? Yet for us, no longer are water, air, or fire the subject of possible "ultimate sources" of everything. These are for us today in the realm of the known elements of physics.

But we must finally now extend and unify our inquiry toward a true ultimate origin, a first causation, the true source of the physical, temporal universe as we know it, without brushing the question aside with some simple strokes of religious jargon or tradition, or some scientific "explanation" of the make-up of physical phenomena.

As we know scientists, from their physical footings, have long been looking for a theory to "unify, explain everything" -- but in vain. The time, therefore, has come for us to go beyond the purely physical – into the world of the invisible, untouchable, the world of pure principles, of universal laws, the realm of the unseen, as of Gravity.

When physicist Einstein conjured up his famous theories, he honed in on, sensed, comprehended and somehow understood the principles making up his theories. He did not actually *know*, or even learn, the practical outcome of his understanding of things until later in life when it was eventually verified in applied physics.

He himself said that the imagination is more important than direct knowledge. We can likewise sense, hone in on – intuitively understand – ultimate consequence in and of all things without a priori empirical knowledge of "the other side," without having been consciously there and back. The whole ultimate picture is quite certainly beyond our direct empirical experience anyway. Our consolation prize is that we have, "on this side," a priori empirical knowledge of the sum of all the alternatives and can find the direction of the truth by deduction, provided we really want to know the truth. Simply put, by finding all things that are wrong with one (our) side, we can closely deduce what is right with the other side.

True total knowledge is and always has been there, complete in and of itself, by virtue of its oneness, its being *synonymous with the complete actual history of all things and events*. It is out there, just as it happened from the first occurrence of a dust particle, of impurity in eternal "space." There is only one way it happened, there is only one way things are the way they are, and therefore can be. The crux of it is that we are just not privy to that total history, to that total knowledge; to what extent we ever will is a matter of the measure of our limitations, perhaps of our will and ultimate capacity to know, to some extent perhaps of conjecture, the imagination, or of the use of deduction by the empirical knowledge at our disposal, perhaps a combination of all these.

What then, in reality, is *true knowledge* but the total history, a true rendering of the complete makeup of all facts, of all things and events from their first origins to their ultimate end? Could it not be said, then, that genuine knowledge is but *known* fact, *known* truth? – Some sort of triad between fact, truth and *genuine knowledge*, "knowledge of that which is really, actually so," seems to emerge here.

But now, then, what if we had this triad complete, if we knew it all, every aspect of total existence and its cause, source and origin, its intent, purpose, its every minute fact? Would we, in our limitation, understand, comprehend what is really going on, and would we want to change things, would we be us, want to be us? If *"they knew,"* would ants want to be ants? Would we want to be ants? I would venture to say that with full comprehension, awareness and understanding on higher levels, we would want to be us no more than we would want to be the ants.

We might just hold free passes to paradise if only we could know and *comprehend* the full facts and truth as they surely exist – if only we could muster understanding to such an unlimited horizon.

In our reality, what do we really seek, want? Ultimate knowledge? That would not necessarily help us. I had just asked the question: "If we *knew everything*, would we still want to be here, or not? Would we really change things?" I sincerely doubt it, for we know quite a bit already, and we have not changed a lot, have we? So knowledge alone does not necessarily answer our plight. What it takes is *ultimate understanding*.

Remember, Einstein did not "know," but he somehow comprehended, understood, before he knew. Just like earlier man did not *know* the earth was round, but somehow comprehended, figured, understood it before he "knew." So yes, we can understand, we can answer the deeper questions without possessing the empirical knowledge – through reasoned comprehension and understanding. We can comprehend without knowing, but knowledge by itself can be meaningless without comprehension and understanding. That is what Confucius must have meant when he said, "When we know, and we know that we know, and when we do not know, we know that we do not know – that is knowledge." But he might just as well be quoted as having said: " . . . that is comprehension, that is understanding."

The trained dog "knows" certain things, but does he comprehend what is really going on? When he knows to perform certain acts, it is not really from "knowledge" in the sense of Confucius' knowledge of comprehension, but merely from certain familiar – non-cognizant – behavior. His natural instincts, of course, are quite another matter, belonging to nature's programming covered elsewhere.

Ultimate, true understanding comes through comprehension, and we acquire it by the recognition, acceptance and absorption of consequential, quintessential fact and truth.

Our one big task, then, is to find that ultimate consequential, quintessential truth. It will point us to the fundamental universal cause and its (physical) effects and lead us in the right direction toward ultimate understanding.

ON THE PROVINCE OF HUMAN JUDGMENT, HORIZON AND PHILOSOPHY

As we search for the fundamental truth, it is philosophy, our highest discipline and reasoning, judging faculty not beholden to preconceptions of any kind, that is eventually destined to find the door to man's deliverance. *It alone* can judge for us rationally what is ultimately appropriate, right, adequate, determinant, consequential and of true meaning.

But in our ultimate enlightenment, it will be Philosophy and Religion together that shall confer in the final fix on these values and cohort toward our practical acceptance, application, implementation and defense of them, as of good versus evil, adequacy or inadequacy, consequence or inconsequence. Either as a result of enlightened understanding between them, or of one triumphing over the other; specifically, Philosophy over Religion – in a similar vane as scientific awareness did at the time of the Inquisition, notwithstanding the sad circumstance that it took the Roman Catholic Church hundreds of years then to finally "allow" the truth and join in with that which, with genuine knowledge, is self-evident – given honest acknowledgment of the true facts and a horizon wide enough to grasp them, in other words, given an appropriate, sincere comprehension level.

As for science, it will remain anchored in the "dust" of the physical realm that is its domain, but there we must leave the eventual undoing, the de-creation, the dissolution of the temporal – (note well: *temporal* so there be no doubt about our proper understanding of this fact) – physical universe, the oblivion of it all – of the stars, the elements, matter, the "bigger picture" factor – to other determinants, other intelligence and/or levels of comprehension, most likely even to a random process and chance. Let's not reject this latter prospect too out of hand, for even our automatic bank-teller machines use the random principle to achieve the very purposeful, precise, close-to-our-heart objective of safely and securely dispensing our daily cash.

Other than in the sole ultimate light and consideration of good or evil, of adequacy or inadequacy, consequence or inconsequence, how could we really presume to ever judge, at best, much more than our own human situation in the maze of universal things and events?

What we *can* very reliably do within the stated purview, though, is to sense, reason out, *comprehend, understand* – in lieu of actually "knowing," or having to "believe in" – the nature of the eventual course of events, final outcome and conclusion, and therefore actually foresee it based on our understanding of highest universal principles. But again, precisely how it will eventually occur, physically, is for us humans – and most likely even for a God – forever behind closed doors. Like the demolition of a building by explosives, we can foretell, even cause the building's collapse, but exactly how it is going to fall, and precisely where every piece of debris will be placed in the end, is impossible to predict; it cannot help but be a matter of chance, of an absolute, utter random process when it occurs.

The final outcome, however, the end result, the purposeful objective (as with the bank-teller machine the dispensing of our exact cash) -- in this case the cessation of temporal and physical manifestation -- is pre-ordained by highest ultimate principles, an outcome inevitable within the latter's self-fulfilling requisites of eternity and infinity.

It would be an exercise in empty futility to try and guess precisely how the physical universe is going to make its exit. That is not important. Instead, it is the invisible, non-physical realm of eternal principles, of eternal law, of the spirit, of original energy as manifested in Gravity – it is, in sum total, only the inevitable, appropriate final outcome that ultimately matters.

It is often said that people involve themselves in "philosophy" because it make them feel important. Does it hold also, then, that the priest in the pulpit is preaching his religion in order to "feel important?" Is it really "feeling important" that makes one judge our religions as not satisfying enough, as not adequate in light of ultimate consequence – leaving one unfulfilled by wanting dogma? Or is it indeed deeper insight and fuller comprehension that is hindering, preventing one's ability to accept, to simply "believe," a scenario handed down by old writings and "traditional" teaching?

Well then, other than science or religion to explain the state of things, where should we turn, for answers to – and proper deductions from – our being, if not to Philosophy? And without "feeling important" about it – as a matter of fact, feeling more humble about it the more one turns to it; actually, in rather exact proportion to the depth of one's philosophical inquiry and understanding.

Philosophy, reasoning, has more to do with inner perception than with outer appearances, loud noises, or with University degrees. It is the inner search, the urge, the need in us toward greater purpose, deeper understanding, real values and ultimate truth; no doubt about that.

True philosophy is undoubtedly the noblest, highest of human endeavors.

There is good reason for us humans to feel inadequate, incomplete, imperfect. And there is something especially incongruent about those studied ones who profess to know all about those light years, and who do not seem to recognize, acknowledge and act upon the inherent shortcomings of creatures living three score and ten in abject dependence, in such obvious inadequacy when compared even with the crudest concepts of perfection.

We can talk about it, can't we? Take a person after a "sumptuous" meal and drink, detain him or her unconditionally for just a few hours, and you approach the dungy "human" problem. The subject is more freely discussed, without a blink of the eye, when on a less personal level, a sewage line by a lake or in a coastal city breaks, when a beautiful natural bay gets contaminated due to sewage spills . . . do you get the point? Very vulnerable, us humans. And – remember our autopsy?

Light years – or better yet, those laws eternal – I much prefer, even without a University diploma.

There is existence out there, you know. We can be sure, for on a clear night you can see plenty of it – go see! And stacked against highest ultimate principles, we can be equally and absolutely sure that the human experience is *by light years far not* the last word, but rather a most humble, barely discernible remnant, or beginning, in enlightened existence.

To need or not to need be

our imperfect, wanting selves –

That is the question.

In full enlightenment,

not to need be thus –

That is the answer.

ON IMPERFECTION AND THE PERFECT STATE

Can we perceive, define, know perfection? Not just a casual, passing kind. Perfection, perfectness as an Absolute?

We certainly can. Even express it mathematically, by deduction: X minus imperfections equals perfection. Which means, take the present, or any state, remove from it all imperfection, and you end up with perfection. As an ideal, as a concept, as a model, certainly. But then, what is "God," if not an ideal, a model?

Sounds simplistic? Not if you have difficulty finding, defining perfection in the absolute any other way, from its outside. If one cannot define the perfect state, at least one would be in a better position to know and identify imperfection (with which we are more familiar). Thus, the perfect state can be approached through alternative deduction, and a fairly clear perception of absolute perfection emerges. As we begin to seriously learn and care to identify imperfection, the task of more adequately perceiving perfection becomes closer at hand.

The definition of imperfection, in light of highest ultimate principles, if honestly pursued, is really self-evident, and the removal, philosophically, of all such imperfection from a given state will point the honest, true seeker clearly in the direction of absolute perfection.

Then, as in a double check, by overlaying those criteria of perceived absolute perfection on our worldly, physical and temporal manifestations, we can easily, and with certainty, judge whether our world and empirical experience here do measure up. If thus we are sincere and care enough about the criteria of ultimate perfection, of perfectness, we shall easily recognize and know our proper direction. To the degree, however, that we philosophically compromise this proper direction, we compromise our relationship to the first origin and ultimate destiny – the perfect state model.

In the deepest, ultimate sense,

there is no other judgment to be made

but over good or evil, as between

perfection and imperfection,

adequacy or inadequacy.

The good, moral, rational and perfect

will surely bear eternal consequence;

evil, the irrational, inadequacy and imperfection

prove inconsequential in eternity.

ON OUR TIME

In the human history, short as it is, there have been varied epochs: of growing awareness, religious awakening, of Black Ages, Renaissance, creation of classical music, of literature, and so on. We are now entering an epoch of truly higher awareness and realization, where old beliefs and superstitions are simply being lost for lack of substance and sufficiency for our new levels of knowledge and understanding.

We must fill the vacuum thus left with a return to basic truth, basic fact and comprehension. True "religion": belief in tangible truth. Genuine knowledge and understanding without "make-beliefs," deception, and self-interests. True teaching.

Values shall be properly placed, population numbers diminish. We are already seeing that happen. The truly more aware, consciously developed peoples are generally partaking less in population explosion than lesser developed ones.

Reason shall prevail and rationality come to be recognized in the good, moral, adequate and perfect; mental, spiritual contentment shall rise as a forerunner to ultimate deliverance from physical bondage. This certainly will not generally happen in the foreseeable or even distant future; in eons of time, ultimately, however, it is the inevitable direction and outcome.

Serious contemplation is needed to get it right. In the young, we have to allow proper, open teaching and education, rather than continuing on with the old, traditional notions that are erroneous and hard to remove once implanted in an infant's heart, soul, and thereby conscience. While more people in our time, especially among the young, are blessed with sufficient true cognizance so as not to be able to accept and believe the old tales and fables told them, where should they turn for proper guidance, for credibility, believability, plausibility and reaffirmation?

Where lies the real truth? Real religion? What is it we really want or should want? Are we accomplishing what we are here to do? Are we, really, ready—getting ready—for higher essential being?

We want to sit down and tangibly pursue it, figure it out once and for all, and deal with it seriously so as to change our lives – meaning our actual essential nature of being. Through serious inquiry and contemplation and a firm will, aspiration and determination to find and foster true cognizance, and to act on the new awareness and *knowing* faith in ultimate, fundamental, essential truth.

In these days' life of "refinements," can you see yourself wanting to live in the rough, backward days "when the West was won," of the Black Ages, of human sacrifices for "God's sake" – albeit "he" often just an idle, blood-thirsty, worldly emperor? Looking back from today's vantage point, those conditions are virtually inconceivable. The answer, therefore, must surely be a resolute NO – for we are a lot more developed now, "much wiser – we know better."

So it will surely be in another time span from now when we shall understand ever more about our lives, about "our" world, the Universe – until we shall, eventually, see the total light and not want to "go back" to ourselves altogether, to be as we are now – *imperfect, wanting, physical in basic nature.*

But to this day, most of us are still simply accepting our situation, particularly with all our gadgets and our lives' comfort gains. Besides, we feel very "superior" compared to our ancestors, while looking yet for better future "gains," for more smarts, and in the midst of all these distractions, we do not seem to have those basic, fundamental, philosophical needs and questions any more. Not like the old classics – not like those monastics who sought real meaning by writing the Bible.

For two to two-and-a-half millennia, indeed not much real, fundamental inquiry has occurred regarding our existence. "Technological, scientific progress" is the new motto and goal, while our technicians feverishly look to develop yet the next, higher-resolution TV screens. But no more real questions. Book stores have all but abandoned the section for "Philosophy."

The really tragic thing is that, as all those teachings on "divinity" date back so many years to when man first became literate and sought meaning in and about life and started making a written record, they have become shrouded in a strange, hard-to-shake mysticism, and we therefore have not much to revere and look to but those old notions, while being "happy" with ourselves and our ever-increasing comforts and hopes, albeit false.

Yes, the time has surely come now when we must reassess our universal standing and seek anew real meaning from our new "plateau" of understanding. That, in a nutshell, is our message, our task.

This is not about guidance on how to live best, on the best social conditions, the best economic system – but rather, it is on our search concerning the fundamental role of universal existence and of our human situation.

Our conclusions touch on our whole being,
and that of the entire Universe,
on very basic existence as we know it.

ON THE TRUE MAKER – DEITY OR IMPOSTOR

Ever since the beginning of human cognizance, there has been one belief or another in a higher authority, a (sympathetic) creator of it all. After all, creation "had to be" – it's all here, isn't it? And if there is a creation, there must have been a creator – *in kind!* So goes conventional teaching and religion.

What if the "creation" of the physical is in fact the work of a creator of another kind, of a "demon" of the negative? The true Deity, the law of ultimate perfection, may just have its hands full keeping all that temporal, physical stuff, the physical universe, in some sort of order with applied gravity, like a dog on a leash.

I am quite sure that gravity is that true, perfect law's energy leash within and on the negative phenomenon of the physical. Some day, when the "dog" dies its final temporal death, this "leash" will haul all physical debris back in for dissolution and re-absorption into the eternal, polarity-less, perfect state. Science has already spoken on such phenomena as black holes absorbing matter – and, conversely, that "It is not inconceivable that the primal energy emerged at zero time from quite literally nothing, uncannily in accord with the structureless singularity described by the time-honored poetic expression: . . . without form, and void, with darkness upon the face of the deep."[1] Moreover, science states: "Though it seems impossible that a particle could materialize from nothing – not even from energy – it so happens that *no laws of physics* are violated . . ."[2] *The violation is, instead, that of the laws of eternity, of infinity, of perfection.*

So it follows that to the enlightened, returning home to the perfect state is what ultimate truth, purpose and understanding is all about.

---

[1] From "Relatively Speaking," E. Chaisson.

[2] Ibid.

ON THE CONDITION OF RELIGIONS

Every once in a while, however rare, we experience "prophets," enlightened persons who see the "other side," or the real, true, invisible essence of things – true, pure principle. They are soon surrounded by peripheral "interpreters" who then build on the backdrop of the true insight but get it distorted, wittingly or unwittingly, and depart from the deep truths for they have not truly comprehended, understood and cognized what their "master" had said and meant. Then a "religion" is created around this periphery of the full, real truth – soon with misguided, self-interested interpretation and teaching at its center.

If it were otherwise, the unselfish truth would, could, should emerge, be taught, pursued, be found, more readily. But being prepared to give things up – power, influence, wealth, even oneself – for the higher principles is something hardly anyone is able or willing to muster without the real, true insight.

"Fanaticism" in the name of religion is nothing new; if true insight were at its core and acted upon, it would be delivering – it would then really no longer be fanaticism, but simple adherence to the reality of cognized truth.

Instead, we keep on thinking and teaching, inverse to real fact and truth, that for the sake of attaining eternal life, we have to start "us" out over and over again on a physical, temporal level. As stated before, eternity, infinity have no end, nor beginning. It is something we have to simply learn to strive for and return to, abandoning our erroneous limited path on this, our plane here.

This is what true religion shall have to come to teach when it attains full insight into the truth, and is ready to act upon it.

ON MAN'S SHAME

Man's shame did not come about through "temptation succumbed to," but with the dawn of his greater awareness, self-cognizance and self-realization – left increasingly wanting in an imperfect setting.

In themselves, physical processes are not out of the ordinary in nature, to which our entire animal kingdom, even our plant life, "shamelessly" attest.

"Temptation" is a metaphoric miscue in a process which rightly has to do with nature's innate, basic imperfection. While it is certain that nature, once set in motion, evolved on its own – set out on its own "program" – what is happening is that we are transcending those natural processes in an increasing spiritual awareness over our wanting, imperfect state of being.

What we particularly perceive as a cause for shame is our manifestly awkward, our "wrong," hush-hush, dishonest way of (re-)creating our imperfect being in the eyes of our truly better, honest judgment and conscience.

In the final analysis, what we aware creatures do with our awareness is what counts. *There* is our tangible opening away from imperfect being and toward our return to perfection. *There* is the true point of reckoning in the practical merits of religion, to a degree of the sciences and ultimately, of course, of philosophy.

ON ETERNITY AND THE TEMPORAL

It seems, *it is* in fact so simple – yet it may be doubted to the bitter end of this (temporal) world: Eternity is – as is Infinity – the true reality, *our* true reality, without beginning and without end, like the principle in a perfect circle (save any empirical conception or form of a circle!).

We continually *talk* about eternal life. We preach, and we sing about it in the churches. It is the center of our hopes in everything. And yet, we still refuse to really stand on the fact that it is the only real, true reality.

We continue to foster and cherish the physical, the temporal, the lost. We are intrigued by time and space. Their antithesis, our real parents – eternity and infinity – the true reality of everything, we consider a fish too hard to catch and forfeit it, despite our avowed cherished hopes and dreams about it.

We turn productive plowshares into destructive, deadly swords. We procreate, we defend in the physical as if it were our salvation. We forfeit the real and true for the brief, meaningless pursuits of the temporal and physical. All along, there is that majestic, magnificent, perfect principle of the eternal, the infinite.

As Gravity, the hand of the perfect, infinite origin, will eventually haul back all physical phenomena, it shall also return to that perfect, eternal origin all temporal by hauling in time – as it will, in fact, have nowhere to go.

Think about it, seriously – reason it out once and for all: *there* lies the rational, reasoned principle, the good, moral, adequate, the true, rational and consequential in all existence. Come home to the eternal. Overcome self, and the temporal, physical worlds.

This philosophical finding, judgment and conclusion is incontrovertible in light of known, cognized, highest universal principles.

Once again, all true, credible religious teaching must eventually come together in harmonious agreement with these fundamental facts and truths.

ON THE NECESSITY AND ESOTERICA OF THE INQUIRY

As stated before, the ancient Greeks' search was more to the point of fundamental inquiry. There wasn't that much technical, modern stuff to worry about and distract them, so with the first real awakening of human cognizance, they were concerned about the basics, like a virgin about virginity. Thales' motto, for example, was to "cognize, become aware of yourself." Eastern thoughts are more in myths, abstract, distracting – which is not to say that there is no truth whatever in them, any more than it would be to maintain that there is no truth to what, for example, the Christian Bible is trying to say. But what should be said is that for the vast human majority, for the multitudes, the real truth is already shrouded enough and hard enough to get to and to comprehend, so that it becomes harmful to the aims of the seeker to make it any more mythical, any more mystical, than need be.

The Middle Ages, through the Renaissance, brought some attempts at revival of fundamental thought, albeit very peripheral (losing itself quickly in old religious anchors), later eventually culminating in the more fundamental thinking of Schopenhauer. Others, virtually to the man, were taking the existing, the given, the here and now, largely for granted as a basis for their inquiry, and their answers ended up in the nature of all but "refinements" and/or "justifications" for it.

Modern thought is just about completely lost in technology, science, and competition. What we desperately need now is a rebirth of fundamental, absolute, to-the-source-of-it-all philosophical thinking. "Unser wahres Wesen zu erkennen," that is our task (to seek self-cognizance and realization of our true essential nature).

There are different levels on which we perceive. It is therefore not possible to make a clear statement on the true nature of things understandable to all alike. Philosophy is, then, by necessity, an elitist, esoteric process. Every creature, and each of us members of the human species, perceives, acts in accordance with the level of its true essential nature.

The fly in the moving railroad dining car – does it know where it is heading? Or is it just preoccupied with the crumbs, the drop of juice there on the tablecloth? Does it really have the capacity to reflect on where it is going? Are not most humans much the same – cruising on the planet earth, around the sun once every 365¼ days and spinning around itself once every 24 hours – not paying any attention to their real journey, instead just preoccupied with the lowly pursuits of their immediate existence and gathering up their "crumbs" – just like the fly in the moving dining car?

It is very rare in the history of our human experience that one of man *seriously, fundamentally contemplates, questions and judges the true nature of things*, the likes of the ancient Greeks or of a Schopenhauer. Instead, man has most always been, at best, just observing, analyzing, investigating, cataloging and speculating on the readily at hand, the *existing* – without much question about the underlying, *fundamental reason or rationale for its being there at all*, while himself merrily going on living nature's course.

Small wonder, then, that after millions of years, it was only just a few short centuries ago that we learned for certain which end of our world and solar system is up – and even in spite of all the new knowledge, have made only minute progress in the old Greeks' pursuit of "what it is truly all about, Alphy."

Unbelievably, it usually takes some doing, even after actual knowledge has been acquired, to make the truth stick – believed and acknowledged – by the old "keepers of the truth," the traditional establishments. Unfortunately, it is much the same today. I may accept my insight and awareness for what it is, true reality, but to prove it to the outside is quite another matter. But never fear, truth works inexorably, and so it will be that enlightened insight shall eventually work itself through to the true nature of things and to our own, imperfect essential nature.

As we said, in only very few instances in millennia have a mere handful among man within all our billions

– aware of, sensing the real "journey" they were on – inquired fundamentally into its apparent mystery, into their essential nature, into their very being. But our multitudes, our untold billions – not unlike most other of nature's creatures – have just been going after their own, busy, limited objectives and physical pursuits. Yes, the little fly in the moving railroad dining car with its little crumbs over and over again . . .

There is a challenge to our proper direction: Make a necessity of the inquiry into ourselves, of the realization *that we are, who we are, where we are,* and to define *our origin and where we are destined,* and *what for.* A necessity it is indeed, if anything we are and do is to have any real meaning and purpose.

Is religion a necessity? It certainly can, or should be, to the human intuition or soul that realizes something is incomplete in a cognizant, aware, knowing existence without a tangible "higher" consequence than just its physical, mortal aspect of "being."

However, let that "religion" not go off on tangents. Let it be a direct, conscious, aware, live *knowing* belief in real truth. What we must come home to is *to believe in that which we know is true, real fact.* Commonly, "belief" is so interconnected with ***not knowing*** that it has no alternative but to simply ask to "just be believed" without any particular foundation in fact. This premise is so very wrong.

Let's believe in that which is real and true, and which is *known* to be so. Why should something not be "believed, believable," simply because it is, in fact, so? Let's remove the wrong notion that if we are to be "believers," we must cling on to something mythical, something mystical, something that is not factual, tangible, real – "that's why we must believe it!" Remember the triad between fact, truth and genuine knowledge? The Inquisitors were so bent on their "belief" that they had to deny the direct, true fact and reality even after they were known. We must not ever let that be repeated.

REFLECTIONS

Over our constant struggle for physical and economic survival, we forget to think about our fundamental being and values.

But then, it has pretty much been that way all along. Early cave man (and animals still) have always instinctively acted and procreated full of preoccupance with security and survival (better with young around, rather than being, and ending up, alone) – same as today, largely, whether it be individually, in groups, or applied to entire species.

This instinct is quite normal, all else being equal and of purpose and rationality. But what if that turns out in the ultimate to be erroneous – at base without good, valid purpose, or rationality? What if our physical life, and all physical existence, turns out to be, in truth, wrong – "sinful," as man's very life source is *in fact* depicted in the Christian Bible – what if the gist of the story of Eve and the Apple turns out to be the quintessential, unqualified spiritual truth winner?

What if the needle's eye passage through the demeaning, secretive, in-the-sewer source to our imperfect being is our constant and final test of consent to our imperfection, of our will to continue on as we are? What then?

You must have guessed by now: I do have news – you can be sure: it is, in fact, so.

In ultimate consequence, the unchangeable laws of infinity and eternity, of perfection, leave no doubt about that. But then again, as related in our "holy book," unawareness is fairly a rational defense against direct responsibility, against guilt; the point being that as long as you are without the awareness, you go among the blind, while nonetheless carrying the burden, bearing the (negative) consequences, of the low life. But once you have gained the higher awareness, you shall bear the responsibility of transcending and being delivered from physical affliction, getting prepared for higher, more perfect existence toward harmony with highest universal principles.

Stepping back a bit – in our own human situation, doesn't it all come down to this single denominator: Man simply believes in, enjoys, his little sand-castle life that (he thinks) is his to control – which he wants to, or in his lack of higher awareness erroneously thinks he needs to, pass on as a means toward reaching his perceived higher purpose. But he thereby in fact misses his homecoming, denies himself his deliverance, the very objective he purports to be pursuing.

* * * * *

We can assume, can't we, that the human species is the most developed of earth's creatures? It has managed to tinker with the very guts of physics, the atom. It is capable, human intelligence, to comprehend the very base of physics' makeup, to ***manipulate it, and to use it for negative purposes***. But now think for a moment, would a perfect God do, intend to do that?

Yes – think about it!

Then, why would a perfect God create infernos like burning, churning stars, have the stuff cool, let little ignorant physical creatures be evolved there to suffer, largely to fight and live off one another, then soon to perish – only to be reborn again, on and on in quite an arbitrary, helter-skelter fashion?

Yes, think about it.

It certainly makes more sense to attribute these physical "accomplishments," all physical manifestation, phenomenon, to a negative force – to be overcome by the aware, conscious, righteous, positive spirit. In the last analysis, that must be our enlightened judgment on the true nature and rationale of our physical universe, as I am sure it is a perfect God's. Our awareness and understanding of true, incontrovertible higher principles tells us that, without any doubt.

Who could state unequivocally that a much greater power (than our own that created the awesome atomic bomb) could not tinker with things on an immensely grander, "universal" scale, creating our physical universe, when it should have left all in its infinite, eternal, perfect state?

The overall negation of the physical state is a philosophical judgment and decision that must be made categorically. But it surely need not mean that we should stop using our motor car or electric energy immediately in today's world because it creates unhealthy fumes, or not eat certain food because it is repugnant to the aware to feed on other creatures. As long as we exist in this, our imperfect state, we have little choice but to live it out as best we can – in worldly humility. Flesh will consume flesh to the point of salvation, of deliverance, from the flesh. You may have heard it before: "It is not important what goes in the mouth, but rather what comes from it."

The important thing is that we take those tangible steps our higher awareness tells us to take, that is to strive for perfection and not to keep re-creating and re-living the imperfect state. That is reasonable, rational, moral, and in tune with our true conscience and with enlightened self-cognizance. That is, in fact, the prevailing of, the conquest by, the enlightened spirit over ourselves, the overcoming of our physical amalgamation, affliction – the deliverance of our higher subconscious (essence) from its present, imperfect, lower energy state.

Voices have been heard. Probably not understood or comprehended, but heard – in some instances over and over, often for millennia. Prophets and philosophers have laid out their cards, sometimes fairly comprehensively – to not much avail. It appears as if the human species, for one reason or another, from its very edge of ultimate understanding and salvation, is putting up the greatest resistance, a wall, against its own deliverance.

It is evident, therefore, that our species does not possess (quite yet) the real self-cognizance that would deliver it, the complete awareness it would take to abandon its crude, imperfect ship for the more perfect good.

We must, therefore, see our purpose, our mission, in gaining and teaching the higher awareness, the self-cognizance and self-realization needed to overcome ourselves and reach purer, higher status.

* * * * *

"Give us your children for the first six years, and we shall keep them forever" is what we must wrest from the prevalent religions. What we need is no more than just to keep their minds *open* so they can make their own decision – not hindered by indoctrination, doubt, fear, intimidation, deceit – but solely by their own free will and deliberation, once they are ready to do so. Not too much to ask for, is it? Perhaps the churches, for their own purposes, borrowed from Socrates who is quoted as saying: "Pay attention to the very young and make the best of them."

Societies have been, for the most part, willing to "hang their hats, their souls," on often vague, even wacky religions for ultimate answers. Now let us for a change, without calling for immediate "scientific verification," hang our hat on a *rational, moral resolve* of our human situation, based on all reliable factors available to us. After all, the men who wrote the now widespread Christian Bible had some marvelous insights without much, if any, scientific knowledge.

But I want to caution you: it is hard to imagine what a sea of protest, of opposition and disgust will come to pass on the part of a species—and its establishments—that has been so thoroughly indoctrinated, that thinks it is so wonderful, "made in the very image of the perfect, infinite, eternal – God himself."

And yet, in purely *physical reality*, when you look closely enough, where are we, essentially, so different from the other species: cat, dog, cattle, ape – you name it, made in the same basic design and structure: legs, head, eyes, nostrils, ears and digestive system? The difference, obviously, is in our mental, spiritual capacity in the direction of processes that are increasingly removed from the (imperfect) physical aspect of the species, *a direction in which we should be looking and going* in pursuit of "likeness to God." Moral, spiritual, infinite, eternal pursuit and rebirth. Not physical, temporal.

We must remove the mystery around the human existence. As there are birds to fly if there was air, fish to swim if there was water, there was bound to be thought and reason emerging, if there was a rational principle in the Universe. If we do not concede, as surely as water will seek to go to the bottom, that when conditions are right for things to occur, the inevitable shall occur under the reigning universal principles, then we are opening ourselves up to all kinds of mysterious theories and belief systems, superstition, even outright hypocrisies. For we are becoming so overwhelmed by the "miracle of it all – of ourselves," that we give up rational, logical thinking and give in to irrational processes.

To be sure, not all is wrong with religion. We have to look at what is right about it, and thrash out the unreal.

* * * * *

ON NATURE AND THE OBJECT WORLD

If our physical being, amalgamation, affliction, is by our own doing, by our own guilt, our sin, why then are there plants, beautiful trees, "life" come about, apparently, *without* "sin" – by nature's inanimate and non-cognizant processes alone? No conscious, causal, "cardinal" sin involved? The Grand Canyon "just happened" – nobody dug it according to an architect's or engineer's plans or specifications . . .

Or did it?

The explanation lies in the bigger scheme of things. We have to go back at large to the original cause, back to the rational, or rather irrational, principle of physical phenomenon. Under ultimate, highest principles, eternity and infinity have it. Whatever occurred and is occurring, and whatever will for whatever reason occur that does not measure up to those highest ultimate principles, *let it alone*, it is not the real thing, true reality. You can be sure of it.

Temporal, physical phenomenon in any form simply is not consequential, other than satisfying, or in itself being a consequence of, some imperfect force, power, act, urge, will, desire. It can only serve innately imperfect, limited purpose and aims, *no matter what the magnitude of the enterprise*, up to and including what we conceive to be the entire physical, temporal universe as we know it.

Enlightened cognizance and accomplished self-realization make that clear beyond any shadow of doubt.

ON THE CATEGORICAL JUDGMENT ON THE NON-COGNIZANT WORLDS

Our categorical judgment on the "bigger picture" of the physical, temporal universe must be one of error, evil, sin and fallacy since we know unequivocally that a perfect God's intent and purpose is infinite and eternal. Even our enduring religions and, foremost, our deepest inner intuitions speak to that, from earliest times. There simply can be no doubt about it. Only forces that want to perpetuate the fallacy of the temporal, of the finite physical, could disagree with these most basic and all-encompassing principles. So it is that the same (negative) force that created the temporal, physical in the first place is still at work re-creating, re-enacting its erroneous "creation."

The molecular processes should not cause us that much puzzlement; once impurity, spacial dust, physical phenomenon, and thereby friction were occurring, those processes evolved consequentially; polarity occurred, and the one-on-one game of positive and negative, attract or reject, of magnetism, of mathematical multiplication principia, was in business.

Our grasp of mathematics can help us out there for better understanding, with its basic principles applied to such foundations of matter and form as molecules, atoms, DNAs and genes, etc. Nature then, simply put, is but practical, applied manifestation of "mathematical principia." Given a certain environment, and conditions, corresponding evolutional effects would occur.

To avoid the suspicion of oversimplification here, take as an illustration our solar system: guided by the universal principle of gravity, all its manifestations, as they occurred, evolved into precise orbits – perfectly balanced, as is the case with the entire Universe.

So it follows that guided by these strong, controlling universal principles, water would go to the bottom, that there would be birds to fly if there was air, and fish to swim if there was water. The mystery diminishes if we observe a fish out of water, or a drowning mouse trying vainly for dry land. Every facet of nature has its proper place, and only in its proper place will it thrive.

To be sure, the process' guiding principles are not "by our bench marks," our standards of reference – further evidence, the kangaroo newborn, a mere couple of inches tiny, blind, finding its pouch alone up the mother's belly for its final development; or that bird, that whale, finding their winter homes far away through the air or under the water, or yet that spider, doing its architectural masterwork. Our own lack of "natural instincts," and our inept vantage point for understanding these phenomena are as much evident as they are irrelevant to those attributes of the natural processes.

A brief synopsis of our "innocent" universal "bigger picture" concerning the object and temporal worlds, then, would go something like this: overall, we have a negative "bigger power" behind the temporal and physical; we have physical phenomena occurring; we have physical "life" evolving, eventually culminating in self-cognizant, self-realizing being that will not only recognize its misplacement in an erroneous, temporal, physical world, but also its true origin – original, eternal live energy – and "long to go home" to it.

So we have, in fact, two negations to affirm in the final resolution of our plight, for our philosophical deliverance: the renunciation of our own physical being, and the categorical negation of all physical phenomena as a whole as being solid manifestations of impurity and imperfection.

This is a most rational outcome and philosophical system, for we know that ultimate consequence – eternity and infinity – *cannot* have anything to do with the temporal and physical as an end in itself.

CONCLUDING REMARKS

Even though, in this age of "higher education," this is a lay effort – the truth, the facts, are solid and intact. Details are often not to be had (as in Zen) – but if you read carefully, you will catch it. *It's all there to put together the true, ultimate perspective about our life, ourselves, and "our" Universe*. There is only one caveat: that you be ready for it -- for seeking, grasping, and accepting the delivering truth.

If I were to offer some directions on just how to go about tangibly reaching one's "Paranirvana," I would have to remind you: there are no specific, concrete directives available. You don't turn left by the third traffic light, and then to the right on the next corner. We are dealing with "understanding of the infinite, eternal" – God, if you will. Invisible, formless – "Thou shalt not make an image of me" still may best explain the proper understanding of these principles. The instant you start or try to visualize this thing through time, form, matter or presence, you lose its heart, its core.

You must start with a good dose of humility about yourself. Take yourself out of the picture. Always keep in mind, always concentrate, meditate on the ultimate goal, objective: the timeless, formless, non-physical – no trivia, no trinkets, only your cherished principles, away from the material, physical domains onto what is eternal, good, adequate, true, moral and rational.

Now, as a concluding thought, I ask:

Which of our known disciplines: Science, Religion – or Hocus-Pocus . . . is in any just, right position to disclaim what we find to be our deepest answers?

Think about it, seriously.

Science has no conscience, and its domain ends with the bounds of the physical, the temporal.

Religion, dogmatized, has its feet off the ground and most of its head in the sand.

How, then, could anything but serious philosophy lead us rationally to the *depths of ultimate fact and truth, to the way things really are?*

May the blessings of full enlightenment and understanding come to you and lead you toward deliverance.

Farewell,

– *SOMOS*